U.S. Department
of Transportation

National Highway
Traffic Safety
Administration

http://www.nhtsa.dot.gov

I0470650

DOT HS 809 719
NHTSA Technical Report

February 2004

Evaluation of FMVSS 214 Side Impact Protection for Light Trucks: Crush Resistance Requirements for Side Doors

1. Report No. DOT HS 809 719	2. Government Accession No.	3. Recipient's Catalog No.
4. Title and Subtitle Evaluation of FMVSS 214 Side Impact Protection for Light Trucks: Crush Resistance Requirements for Side Doors		5. Report Date February 2004
		6. Performing Organization Code
7. Author(s) Marie C. Walz		8. Performing Organization Report No.
9. Performing Organization Name and Address Evaluation Division; Planning, Evaluation and Budget National Highway Traffic Safety Administration Washington, DC 20590		10. Work Unit No. (TRAIS)
		11. Contract or Grant No.
12. Sponsoring Agency Name and Address Department of Transportation National Highway Traffic Safety Administration Washington, DC 20590		13. Type of Report and Period Covered NHTSA Technical Report
		14. Sponsoring Agency Code
15. Supplementary Notes		

16. Abstract

Beginning September 1, 1993, all light trucks (pickup trucks, vans, and sport utility vehicles) were required to meet a crush resistance standard for side doors. Data from calendar years 1989 through 2001 of the Fatality Analysis Reporting System (FARS) were used to determine the effectiveness of changes made by vehicle manufacturers to meet this standard. Effectiveness was determined by comparing changes in the number of fatalities in side impacts relative to those in frontal impacts.

Three analysis techniques were applied to the data. First, simple ratios of side-impact to frontal crash fatalities were computed, with comparisons made between vehicles with and without side door beams. Second, side impact fatality rates per one thousand vehicle registration years were determined, with vehicles separated according to whether they were manufactured before or after side door beam installation. Finally, a regression analysis of the ratio of side-impacts to frontal fatalities as a function of the presence of side door beams was done.

The effectiveness of side door beams for front outboard occupants was estimated to be 19 percent in all single vehicle side impacts, which would result in the saving of 151 lives in those type crashes if all light trucks were equipped with the side beams. Looking at single vehicle *nearside* impacts only, the effectiveness of the beams was estimated to be 25 percent. If all light trucks were equipped with side door beams, an estimated 124 lives would be saved annually in single vehicle nearside impacts. Little or no effectiveness was found in multi-vehicle crashes.

17. Key Words FMVSS 214; side impact; side door beam; light trucks; crush resistance	18. Distribution Statement Document is available to the public through the National Technical Information Service, Springfield, Virginia 22161		
19. Security Classif. (Of this report) Unclassified	20. Security Classif. (Of this page) Unclassified	21. No. of Pages 45	22. Price

Form DOT F 1700.7 (8-72)   Reproduction of completed page authorized

TABLE OF CONTENTS

Executive Summary

From January 1, 1973 until August 31, 1993, Federal Motor Vehicle Safety Standard (FMVSS) 214 had been applicable to passenger cars only. Beginning September 1, 1993, all light trucks (pickup trucks, vans, and sport utility vehicles) were also required to meet the crush resistance standard, which specified minimum crush resistance when a load is applied to the outer surface of a vehicle door. Manufacturers were permitted to meet the standard earlier, and in a few cases this was done, but the majority of light trucks were installed with side door beams beginning in model year 1994. Manufacturers met the static crush requirement of FMVSS 214 by installing longitudinal side door beams.

Data from calendar years 1989 through 2001 of the Fatality Analysis Reporting System (FARS) were used. Vehicle model years used in the analysis ranged from 1989 through 1997. These data included 1,376 cases of single vehicle side impact fatalities. This is the type of crash most likely to benefit from the presence of the side door beams.

An earlier evaluation on the effectiveness of side door beams in passenger cars had found the greatest benefit to be in single vehicle side impacts, and little or no effect on fatalities in multi-vehicle crashes. In the current evaluation, side impact data were first compared to frontal impacts, which served as a control group. Effectiveness of installing side door beams was determined by comparing changes in the number of fatalities in side impacts relative to those in frontal impacts. Data were examined for all front outboard occupants, as well as for drivers and right front passengers individually. Vehicles were limited to those without changes in the presence of air bags, or analytic techniques were used to control for air bags, since this would clearly have an impact on the number of fatalities in frontal impacts.

Three analysis techniques were applied to the data. First, simple ratios of side-impact to frontal crash fatalities were computed, with comparisons made between vehicles with and without side door beams. Second, side impact fatality rates per one thousand vehicle registration years were determined, with vehicles separated according to whether they were manufactured before or after side door beam installation. Finally, a regression analysis of the ratio of side-impacts to frontal fatalities as a function of the presence of side door beams was done. The regression analysis allowed control of additional influencing variables, such as the presence of air bags and vehicle age.

The regression analysis was considered the best technique, because it allowed several factors to be investigated at the same time, such as seat position, presence of air bag, and vehicle age, which could have affected the previous two analyses, as well as allowing a larger sample size, since data before and after beam installation did not have to be matched for such situations as presence of air bags or years of production.

Regression models were run for various combinations for side impacts. All side impacts were combined; single and multi-vehicle crashes were also examined separately, as were near and far side crashes. Presence of the side door beam was found to statistically significantly reduce fatalities in single vehicle side and single vehicle nearside impacts, relative to frontal crashes, for drivers alone as well as in combination with right front passengers. The regression analysis showed that side door beams are effective in preventing front outboard fatalities in single vehicle side impacts. There was a

19 percent reduction in fatalities attributed to the beams for both drivers alone as well as for all front outboard occupants. In single vehicle nearside impacts, drivers alone saw a 26 percent reduction, while the reduction for all front outboard occupants was 25 percent. All four of these reductions were statistically significant. Right front passengers also saw sizable 18 and 17 percent reductions, in single vehicle and single vehicle nearside impacts, respectively, although these were not statistically significant. Little or no effectiveness was found in multi-vehicle crashes.

The effectiveness of side door beams for front outboard occupants was estimated to be 19 percent in all single vehicle side impacts, and 25 percent in single vehicle nearside crashes. Based on these two different effectiveness estimates, the number of lives saved annually was estimated. Using the 19 percent fatality reduction for front outboard occupants in single vehicle side impacts, it is estimated that, if all light trucks were equipped with side door beams, 151 lives would be saved each year in single vehicle side impacts (both nearside and far side). The 95 percent confidence band for effectiveness for all front outboard occupant single vehicle side impact fatalities ranged from 4 to 32 percent. The 95 percent confidence bounds range from 29 to 285 lives saved if all light trucks were equipped with side door beams. These calculations are based on the average effect of side door beams over all single vehicle side impact fatalities, and applying the effectiveness estimate to the total of nearside and far side fatalities combined.

A slightly more conservative estimate can be obtained by assuming the beams are effective only in nearside single-vehicle crashes, and have little effect in the far side crashes. Using the 25 percent effectiveness estimate for all front outboard occupants in single vehicle nearside fatalities, the 95 percent confidence band for all front outboard occupant nearside fatalities ranged from 8 to 39 percent. If all light trucks were equipped with side door beams, an estimated 124 lives would be saved annually in single vehicle nearside impacts. The 95 percent confidence band is 35 to 222 lives saved.

Introduction and Background

The National Highway Traffic Safety Administration (NHTSA) issues Federal Motor Vehicle Safety Standards (FMVSS) that must be met by vehicles manufactured for sale in the United States. FMVSS 214 ("Side Impact Protection") specifies performance requirements for the protection of occupants in side impact crashes, in order to reduce the risk of serious and fatal injury to occupants. A primary objective of FMVSS 214 is to minimize danger caused by intrusion into the passenger compartment. FMVSS 214 has both a crush resistance requirement for side doors as well as a side impact requirement based on dynamic testing. This report covers only the extension of the crush resistance requirement to light trucks.

The initial version of FMVSS 214 was limited to a crush resistance requirement for passenger cars, effective January 1, 1973. On December 22, 1989, NHTSA published a Notice of Proposed Rulemaking (NPRM) to extend FMVSS 214's existing passenger cars quasi-static test requirement to light trucks. This would, in effect, require light trucks to be equipped with side door beams.

From January 1, 1973 until August 31, 1993, FMVSS 214 had been applicable only to passenger cars. In most cases, manufacturers met FMVSS 214 by equipping cars with a longitudinal beam in vehicle doors. NHTSA evaluated the benefits of the crush resistance regulation in passenger cars, and found single vehicle side impact occupant fatalities were reduced by 14 percent, saving 480 lives annually (Kahane, 1982).

FMVSS 214 was extended to light trucks in 1991 (for the final rule extending static test to light trucks, see *Federal Register* 56 (June 14, 1991): 27427), with an effective date of September 1, 1993. Beginning with model year 1994, all light trucks met the standard. Some vehicles were already being manufactured with side door beams prior to the effective date of the standard.

The 1982 evaluation of the crush resistance requirement found the greatest benefit was for passenger cars in single vehicle side impacts. The crush resistance requirement was found to have little or no effect on reducing fatalities in multi-vehicle side impacts. Thus, the focus of this report is on single vehicle crashes, although data from multi-vehicle crashes are presented for comparison. In addition, occupant seating position is noted, to distinguish between nearside and far side impacts.

The crush resistance test consists of gradually forcing a steel cylinder of 12 inch diameter into the door. With seats removed, the cylinder must encounter a resistance averaging at least 2,250 pounds during the first 6 inches of crush, averaging at least 3,500 pounds during the first 12 inches, and reaching a peak of at least 7,000 pounds or twice the vehicle's curb weight (whichever is less) at some point during the first 18 inches of crush. With seats installed in the vehicle, the cylinder must encounter a resistance averaging at least 2,250 pounds during the first 6 inches of crush, averaging at least 4,375 pounds during the first 12 inches, and reaching a peak of at least 12,000 pounds or three and one half times the vehicle's curb weight (whichever is less) at some point during the first 18 inches of crush.

Analysis Approaches

The Fatality Analysis Reporting System (FARS) is a census of fatal traffic crashes since 1975. Information on the crash as well as the vehicles and people involved is recorded in the data file. For these analyses, front outboard (drivers and right front passenger) fatalities were selected from the FARS file. This report focuses on the effectiveness of the amended standard in preventing fatalities only.

This analysis examines fatality data in a number of ways. Initially, side impacts are grouped according to the number of vehicles in the collision (single or multi-vehicle) as well as whether the impact occurred on the side where the occupant was seated ("nearside") or on the opposite side of the vehicle ("far side"). The numbers of fatalities in the specific type of crashes are compared before and after the trucks were equipped with side door beams. Frontal impacts are used as a comparison group, to determine whether any changes observed in side impact fatalities were due to the addition of side door beams or to other, unrelated causes. Pure frontal impacts only were used, to eliminate contamination with frontal impacts also containing a substantial side impact component. Although this reduces the size of the comparison group, it allows for a cleaner comparison.

Registration data were also used to determine fatality rates before and after the installation of side door beams in light trucks. This allowed the evaluation of the beams' effectiveness without having to rely on a comparison group, which might also have been changing in unknown but systematic ways, either similar or dissimilar to the beams' effect.

Finally, a regression analysis was performed, which permitted the control of several possible factors. Data from both side and frontal impacts were included, and the impact of other variables on the type of crash was measured. This method allowed factors previously controlled for in other ways in the two earlier analyses to be a more integrated part of the analysis. For example, rather than having to eliminate vehicles with alternate air bag status, the presence or absence of such could simply be noted in the data and would be incorporated into the analysis. This resulted in a larger sample size as well as a more statistically sophisticated method. The results of the regression analysis may not be as easily comprehended as, for example, the difference in fatality rates for side impacts as compared to frontals. However, the results of all three analyses were consistent, showing a reduction in single vehicle side impact fatalities after side door beams were installed in light trucks.

Data Preparation

Examination of the number of fatalities and crashes before and after vehicles met the new standard provided valuable information. Specifically, "Before" and "After" groups of certain vehicles having no substantive structural or safety related changes other than the addition of side door beams (or other changes in the side structure) were compared to quantify changes in crash rates, as well as the types of crashes occurring.

If manufacturers had phased in the necessary changes, they would have been required to report specific information to NHTSA, including when each make-model was first equipped with side door beams or other modifications to meet FMVSS 214. However, the requirement was not phased in, and therefore, the information was not reported to NHTSA. The manufacturers then had to be contacted for information on when specific vehicles met the standard for light trucks.

Representatives from Daimler-Chrysler, Ford, General Motors, Nissan, and Toyota provided information on when their light trucks first began meeting the new standard. Exhibit 1 presents the vehicle data used for this report. Vehicle clones, essentially twins built on the same platform, are considered the same vehicle model throughout this report. Vehicles are combined into groups for simplicity where there is no difference in the model years used for the current analysis. In cases where alterations in the vehicle resulted in differences in the usable model years for the analysis, the vehicles are listed separately.

If too long a span of years were to be used for the before/after comparisons, observed differences could potentially be due to a comparison of older to newer vehicles. Therefore, vehicles no more than four years before and after the installation of side door beams were used. If four model years of data were valid in both the "Before" and "After" group for the vehicle in question, all those data would be used. If either the "Before" or "After" group was limited to a shorter time span, due to air bag installation, model redesign, or non-production, then both groups would use the shorter time span. The same number of years was used for each of the two groups, with and without side door beams, for each make-model. The columns labeled "1st Year Produced" and "Last Year Produced" refer to the model years of production for that specific vehicle during which no major redesign took place.

The column "First Year with Beams" notes the model year that the vehicle implemented changes in the side structure to meet the extension of FMVSS 214. The majority of light trucks made these changes in model year 1994. The exceptions are:

- Ford Econolines were redesigned in 1992 and had beams beginning then
- Ford Explorers began production in 1991 and have always had beams
- Chevy Lumina, Pontiac Transport, and Oldsmobile Silhouette began production in 1991 and have always had beams
- Nissan Pathfinder and Nissan Frontier first had beams in MY 1993
- Nissan Quest began production in 1993 and have always had beams
- Toyota Previa began production in 1991 and have always had beams
- Toyota LandCruiser had beams beginning 1993
- Mazda Navajo began production in 1991 and have always had beams

Only the Nissan trucks and Toyota LandCrusiers listed above could be used for this report, as the other vehicles met the standard either when first produced or when redesigned, preventing any "Before" data.

Air bags are effective at preventing fatalities in frontal impacts. Because frontal impacts are used as the control group, it was also important to note the presence of air bags, in order to maintain consistency in the "Before" and "After" groups. The columns labeled "First Year Driver AB" and "First Year RFP AB" provide information on the model year air bags were first installed in the vehicle at the driver and right front passenger position, respectively. Entries of '1' in these columns signify vehicles that did not get air bags within four years of either meeting FMVSS 214 and/or a

Exhibit 1: Light Truck Safety Equipment by Model Year

Make/Model	1st Year Produced	Last Year Produced	1st Year with Beams	1st Year Driver AB	1st Year RFP AB	Drivers MY without Beam	Drivers MY with Beam	RFP MY without Beam	RFP MY with Beam
Dodge RamVan	85	97	94	95	(1)	93	94	90-93	94-97
Dodge Dakota	87/90	96	94	94	(1)	none	none	91-93	94-96
Dodge Caravan Plymouth Voyager	91	95	94	91/92	94	92-93	94-95	none	none
Dodge Grand Caravan Plymouth Grand Voyager Chrysler Town & Country	91	95	94	92	94	92-93	94-95	none	none
Jeep Grand Cherokee	93	01+	94	93	96	93	94	93	94
Ford F150 Pickup	85	96	94	94	(1)	none	none	91-93	94-96
Ford F250/350	85	97	94	(2)	97	none	none	91-93	94-96
Ford F250/350 (SuperCab, Crew Cab)	85/89	97	94	(1)	(1)	90-93	94-97	90-93	94-97
Ford Aerostar	86	97	94	92	(1)	92-93	94-95	90-93	94-97
Ford Ranger	93	97	94	95	95	93	94	93	94
Mercury Villager	93	98	94	94	96	none	none	93	94
Chevrolet Astro GMC Safari	85	N/A	94	94	96	none	none	92-93	94-95
Chevrolet S10 Blazer 2D GMC S15 Jimmy 2D	85	01+	94	95	(1)	93	94	90-93	94-97
Chevrolet S10 Blazer 4D GMC S15 Jimmy 4D Oldsmobile Bravada 4D	91	01+	94	95	(1)	93	94	91-93	94-96
Chevrolet K1500 4x4 Blazer, Tahoe GMC Yukon	92	97+	94	95	97	93	94	92-93	94-95
Chevrolet C/K Pickup GMC Sierra Pickup (a)	88/92	N/A	94	(2)	(1)	93	94	93	94
Chevrolet C/K Pickup GMC Sierra Pickup (b)	88	N/A	94	95	(1)	90-93	94-97	90-93	94-97
Chevrolet C/K Pickup GMC Sierra Pickup (c)	88	98+	94	95	97	93	94	91-93	94-96
Chevrolet C/K Pickup GMC Sierra Pickup Crew Cab (b)	92	N/A	94	(1)	(1)	92-93	94-95	92-93	94-95

Make/Model	1st Year Produced	Last Year Produced	1st Year with Beams	1st Year Driver AB	1st Year RFP AB	Drivers MY without Beam	Drivers MY with Beam	RFP MY without Beam	RFP MY with Beam
Chevrolet C/K Pickup GMC Sierra Pickup Crew Cab (c)	92	N/A	94	95	97	93	94	92-93	94-95
Chevrolet S/T Pickup GMC S/T Sonoma	88	N/A	94	95	(1)	93	94	90-93	94-97
Chevrolet G10 & G20 ChevyVan, G10 SportVan, GMC G15 Rally, G15 & G25 Vandura	85	95	94	94	(1)	none	none	92-93	94-95
Chevrolet G30 ChevyVan, G20 & G30 SportVan, GMC G25 & G35 Rally, G35 Vandura (a)	85	95	94	(1)	(1)	92-93	94-95	92-93	94-95
Chevrolet G30 ChevyVan, G20 & G30 SportVan, GMC G25 & G35 Rally, G35 Vandura (c)	85	95	94	94	(1)	none	none	92-93	94-95
Chevrolet G30 ChevyVan, G30 SportVan G35 Rally, G35 Vandura - extended	90	95	94	(2)	(1)	none	none	92-93	94-95
Chevrolet Suburban GMC Suburban	92	99	94	95	97	93	94	92-93	94-95
Nissan Pathfinder	90	95	93	96	(1)	90-92	93-95	90-92	93-95
Nissan Pickup	85/86	97	93	96	(1)	90-92	93-95	89-92	93-96
Toyota 4Runner	85	95	94	(1)	(1)	92-93	94-95	92-93	94-95
Toyota LandCruiser	91	N/A	93	95	96	91-92	93-94	91-92	93-94
Toyota Pickup	85	95	94	(1)	(1)	92-93	94-95	92-93	94-95
Toyota T100 Pickup	93	95/96/98	94	94	(1)	none	none	93	94

(a) VIN4=G   (b) VIN4=H,J,K   (c) Remaining VIN

1 = Did not get air bags within 4 years of beam installation and/or before major redesign   2=Undetermined

major redesign. Thus, during the model years used for this analysis, these vehicles did not get air bags in the noted seat position. Entries of '2' in these columns indicate that the initial year of an air bag in that seat position could not be determined. Typically, this meant that air bags were optional, and thus the presence or absence in a specific vehicle was unknown. These vehicles were eliminated beginning with the model year in which the presence of an air bag in a specific vehicle could not definitively be established.

The last four columns of Exhibit 1 present the actual model years of each vehicle used in the analysis, for drivers and for right front passengers, each with and without the side door beams. The remaining columns (initial model years of beam installation, driver and right front passenger air bags, and production), determine which model years could be used for analysis. For example, for the Dodge Grand Caravan and similar vehicles (Plymouth Grand Voyager, Chrysler Town & Country), beams were introduced in model year 1994. Driver air bags were first installed in 1992, and dual air bags became standard in 1994. These vehicles were produced without a redesign from 1991 through 1995, three years before the beams were installed and two years after. Thus, two years of "With Beam" and "Without Beam" data could be used, in order to keep the number of years before and after the same. However, since passenger air bags were first installed in the same model years as beams, no passenger data were usable. Thus, for drivers, the "Without Beam" or "Before" group is comprised of model years 1992 and 1993, and the "After" years 1994 and 1995. For right front passengers, no data can be used for these vehicles, so "none" appears in the appropriate columns.

For this report, if a vehicle underwent a major redesign, it was considered a different vehicle. This prevented differences due to structural changes from being attributed to the installation of side door beams. In some cases, more than one model year is listed as the first and/or last year produced. In such cases, some of the vehicles in question were manufactured over a slightly different span of model years.

For example, Chevrolet C/K and GMC Sierra pickup trucks are similar enough to be considered the same vehicle. Because of differences within the model, however, they are listed in five groups in Exhibit 1. In model years 1995 and 1996, vehicles with a Gross Vehicle Weight Rating (GVWR) over 8,500 pounds in this group did not have air bags, while the lighter vehicles had driver air bags. In 1997, the lighter vehicles had dual air bags installed. The GVWR can be determined by the VIN in many cases, but not all. Thus, one group is known to have either driver or dual air bags in these model years (noted as (c) in the Exhibit), another is known to not have any (noted as (b)), and for the third group the presence of air bags is unknown (group (a)). For this third group, only model years 1993 (as "Before") and 1994 (as "After") can be used for both drivers and right front passengers, since beginning in 1995, air bag status is unknown. The same model years are usable for drivers in the group getting air bags in 1995 (c), since that is the model year the change is known. For right front passengers in this group, model years 1994 through 1996 can serve as the "After" years, since no change took place until 1997 for that seating position. For the group that did not have air bags installed until *after* 1997 (b), the full range of model years can be used (1990-1993 as "Before" and 1994-1997 as "After") since air bags are absent in all these years. Production of C/K and Sierra regular cabs began production in 1988, and the crew cab in 1992. Note that for one of the groups (a), in the "1st Year Produced" column, the entry for two of the groups reads "88/92". For this group, the difference in production start does not affect the model years that can be used in this analysis, so they are combined for simplicity. This is not true for the other groups of C/K and Sierra pickups. Since the Crew Cabs were not produced until 1992, there are only two years of "Before" data available, at most, which would then allow "After" data only through 1995 at the latest.

Some light trucks manufactured prior to model year 1993, such as the Ford Explorer and Pontiac TranSport, had side beams installed since they were first manufactured. Since there was no "Before" group available for comparison, these vehicles could not be included in the analysis. Other vehicles that did not begin production until model year 1994 or later, of course, had also always met the side impact standard, and were also excluded from the analysis.

Effectiveness of Side Door Beams in Preventing Fatalities

Since FARS contains only fatal crashes, it is not possible to determine fatality rates per 100 crash involved occupants. However, the data can be used to indirectly determine the relative fatality risk of pre- and post-standard light trucks. This was done by comparing occupant fatalities in side impacts to a control group not affected by FMVSS 214. These groups should be as similar as possible except for whatever effect FMVSS 214 would have. Since door beams were expected to improve vehicle safety in side impacts, and have no effect on purely frontal impacts, any improvements seen in side impacts relative to the frontal impacts would be attributable to the presence of side door beams. Frontal impacts are used as the control group in this analysis. Air bags were the only substantial safety change affecting frontal crashes that was implemented simultaneous with, or close to the installation of beams. As discussed in the preceding section, data were limited to model year ranges in which make-models either (1) were never equipped with air bags or (2) were always equipped with air bags.

Note that, to keep up to four model years before and after the side door beams were installed, model years 1989 through 1997 would be used. Model year 1989 would apply only to those vehicles first getting side door beams in 1993 (e.g., Nissan pickup trucks). FARS data from calendar years 1989 through 2001 were used.

In the analyses below, frontal impacts are used as the control group. Specifically, pure frontal impacts were used, meaning only those impacts at the direct 12 o'clock position. Impacts at the 11 and 1 o'clock position, while typically considered frontal impacts, would have a side impact component to them as well, which could be affected by the installation of side door beams. Therefore, such crashes were not included in the frontal/control group. Side impacts were those in the 2, 3, 4, 8, 9 and 10 o'clock positions.

Generally, separate analyses were performed for single vehicle and multi-vehicle crashes. Single vehicle frontal impacts serve as the control for single vehicle side impacts, and multi-vehicle frontals for multi-vehicle side. Crashes were determined to be either single or multi-vehicle crashes based on the number of vehicle forms recorded in FARS. In addition, data were looked at by the location of the occupant in relation to which side of the truck was struck. Thus, analyses are performed separately for nearside and far side occupants, as well as for all occupants combined. In addition, tables show drivers and right front passengers separately, as well as all front outboard occupants combined.

Data are presented in four sets, from one to four years before and after beam installation. It would be preferable to limit the number of years examined, so that all vehicles in the analysis are as similar as possible, including age of vehicle. However, the additional usable data obtained by including vehicles up to four years before and after beam installation is worth examining.

Using the Dodge Ram Van, the first row in Exhibit 1, as an example, the same vehicle was produced from 1985 through 1997. Thus, 1997 would be the last model year usable for this vehicle. Side door beams were first installed in 1994, which would then be the first "After" year. Any model years previous to this would be "Before." Driver air bags were first installed in 1995, while front passenger air bags were not installed during the production of this vehicle (that is, at least before a major redesign). Since the first year of the "After" group was 1994, and air bags were installed for drivers in 1995, only one year of data for the "After" group was usable for drivers. For the first two analyses that follow (effectiveness based on Before vs After counts and changes in fatality rates), the same number of model years must be used for both the before and after groups. Therefore, for drivers for the Dodge Ram Van, 1993 serves as the only "Before" year, while 1994 serves as "After." For right front passengers, the addition of air bags does not effect the usable model years. Therefore, data as late as the last year of production, 1997, were usable. The "After" years for right front passengers were then 1994 through 1997. The associated "Before" years were 1990 through 1993. Four years before and after side beam installation were the maximum span used, in an effort to control vehicle age from influencing the outcome.

When data appear in later tables as "One year before and after," all data within one year of side beam installation are included. Since all vehicles would require at least one "Before" and one "After" model year, this would necessarily include all models in the analysis. Columns labeled "Two/Three/Four years before and after" include data within the stated time span. For example, "Two years before and after" includes anything within two years of side beam installation. This includes all counts in the "One year before and after" group as well as data one model year beyond that in each direction (both "Before" and "After). Each successive number of years includes the previous smaller group(s).

Data are separated into mutually exclusive crash types for side and frontal crashes. In some cases, occupants are separated into nearside or farside, in relation to the side of the vehicle on which the impact took place. Thus, a driver is a farside occupant when the impact is at the 2, 3, or 4 o'clock position (the passenger side), and a nearside occupant when the impact is at the 8, 9, or 10 o'clock position.

The previously described data were used to compare numbers of fatalities before side door beams were installed in light trucks to those vehicles with beams. Exhibit 2 presents data from FARS comparing single vehicle side and frontal crashes, and multi-vehicle side and frontal crashes. Each data point represents a front outboard fatality – data are presented first for all frontboard occupants combined (Exhibit 2a), and then separately for drivers (Exhibit 2b) and right front passengers (Exhibit 2c). The four pairs of columns contain data on vehicles one through four years before and after the introduction of side door beams. Each pair of columns ("No Beam" and "Beam") shows the number of front outboard occupants with and without the

Exhibit 2a: Front Outboard Occupant Fatalities, Side vs Frontal Comparisons,

Crash Type	One year before and after		Two years before and after		Three years before and after		Four years before and after	
	No Beam	Beam	No Beam	Beam	No Beam	Beam	No Beam	Beam
(1) SV NearSide	123	102	173	142	214	161	232	169
(2) SV FarSide	90	76	118	103	137	114	147	119
(3) SV Pure Frontal	396	380	557	492	665	569	699	587
(4) MV NearSide	212	218	350	300	423	341	442	360
(5) MV FarSide	118	105	179	138	213	161	222	166
(6) MV Pure Frontal	681	588	989	806	1,167	939	1,244	977
Total	1,620	1,469	2,366	1,981	2,819	2,285	2,986	2,378
All Side (1+2+4+5)	543	501	820	683	987	777	1043	814
Frontal (3+6)	1077	968	1546	1298	1832	1508	1943	1564
Effectiveness	-3%		1%		4%		3%	
Chi-Square	0.12		0.02		0.57		0.29	
NearSide (1+4)	335	320	523	442	637	502	674	529
Frontal (3+6)	1077	968	1546	1298	1832	1508	1943	1564
Effectiveness	-6%		-1%		4%		2%	
Chi-Square	0.46		0.01		0.40		0.14	
Far Side (2+5)	208	181	297	241	350	275	369	285
Frontal (3+6)	1077	968	1546	1298	1832	1508	1943	1564
Effectiveness	3%		3%		5%		4%	
Chi-Square	0.09		0.13		0.28		0.23	
SV side (1+2)	213	178	291	245	351	275	379	288
SV Frontal (3)	396	380	557	492	665	569	699	587
Effectiveness	13%		5%		8%		10%	
Chi-Square	1.24		0.20		0.80		1.08	

Exhibit 2a continued: Front Outboard Occupant Fatalities, Side vs Frontal Comparisons,

Crash Type	One year before and after		Two years before and after		Three years before and after		Four years before and after	
	No Beam	Beam	No Beam	Beam	No Beam	Beam	No Beam	Beam
SV NearSide (1)	123	102	173	142	214	161	232	169
SV Frontal (3)	396	380	557	492	665	569	699	587
Effectiveness	**14%**		**7%**		**12%**		**13%**	
Chi-Square	0.92		0.32		1.17		1.51	
SV Far Side (2)	90	76	118	103	137	114	119	
SV Frontal (3)	396	380	557	492	665	569	699	587
Effectiveness	**12%**		**1%**		**3%**		**4%**	
Chi-Square	0.56		0.01		0.04		0.07	
MV Side (4+5)	330	323	529	438	636	502	664	526
MV Frontal (6)	681	588	989	806	1167	939	1244	977
Effectiveness	**-13%**		**-2%**		**2%**		**-1%**	
Chi-Square	1.69		0.04		0.07		0.01	
MV Nearside (4)	212	218	350	300	423	341	442	360
MV Frontal (6)	681	588	989	806	1167	939	1244	977
Effectiveness	**-19%**		**-5%**		**0%**		**-4%**	
Chi-Square	2.45		0.30		0.00		0.19	
MV Far Side (5)	118	105	179	138	213	161	222	166
MV Frontal (6)	681	588	989	806	1167	939	1244	977
Effectiveness	**-3%**		**5%**		**6%**		**5%**	
Chi-Square	0.04		0.20		0.30		0.20	

Exhibit 2b: Driver Fatalities Only, Side vs Frontal Comparisons, One to Four Years Before and After Beams Installed

Crash Type	One year before and after		Two years before and after		Three years before and after		Four years before and after	
	No Beam	Beam	No Beam	Beam	No Beam	Beam	No Beam	Beam
(1) SV NearSide	87	65	112	82	124	89	128	90
(2) SV FarSide	76	67	97	82	104	88	110	92
(3) SV Pure Frontal	319	301	418	359	462	414	474	422
(4) MV NearSide	165	161	246	203	265	222	271	228
(5) MV FarSide	84	81	117	102	127	111	128	114
(6) MV Pure Frontal	538	460	722	569	782	640	811	661
Total	1,269	1,135	1,712	1,397	1,864	1,564	1,922	1,607
All Side (1+2+4+5)	412	374	572	469	620	510	637	524
Frontal (3+6)	857	761	1140	928	1244	1054	1285	1083
Effectiveness	-2%		-1%		3%		2%	
Chi-Square	0.06		0.01		0.16		0.11	
Nearside (1+4)	252	226	358	285	389	311	399	318
Frontal (3+6)	857	761	1140	928	1244	1054	1285	1083
Effectiveness	-1%		2%		6%		5%	
Chi-Square	0.01		0.06		0.45		0.42	
Far Side (2+5)	160	148	214	184	231	199	238	206
Frontal (3+6)	857	761	1140	928	1244	1054	1285	1083
Effectiveness	-4%		-6%		-2%		-3%	
Chi-Square	0.11		0.25		0.02		0.07	
SV side (1+2)	163	132	209	164	228	177	238	182
SV Frontal (3)	319	301	418	359	462	414	474	422
Effectiveness	14%		9%		13%		14%	
Chi-Square	1.16		0.51		1.41		1.63	

13

Exhibit 2b continued: Driver Fatalities, Side vs Frontal Comparisons,

Crash Type	One year before and after		Two years before and after		Three years before and after		Four years before and after	
	No Beam	Beam	No Beam	Beam	No Beam	Beam	No Beam	Beam
SV Nearside (1)	87	65	112	82	124	89	128	90
SV Frontal (3)	319	301	418	359	462	414	474	422
Effectiveness	**21%**		**15%**		**20%**		**21%**	
Chi-Square	1.64		0.97		2.07		2.39	
SV Far Side (2)	76	67	97	82	104	82	110	92
SV Frontal (3)	319	301	418	359	462	414	474	422
Effectiveness	**7%**		**2%**		**6%**		**6%**	
Chi-Square	0.13		0.01		0.13		0.16	
MV Side (4+5)	249	242	363	305	392	333	399	342
MV Frontal (6)	538	460	722	569	782	640	811	661
Effectiveness	**-14%**		**-7%**		**-4%**		**-5%**	
Chi-Square	1.35		0.45		0.17		0.31	
MV Nearside (4)	165	161	246	203	265	222	271	228
MV Frontal (6)	538	460	722	569	782	640	811	661
Effectiveness	**-14%**		**-5%**		**-2%**		**-3%**	
Chi-Square	1.07		0.17		0.05		0.09	
MV Far Side (5)	84	81	117	102	127	111	128	114
MV Frontal (6)	538	460	722	569	782	640	811	661
Effectiveness	**-13%**		**-11%**		**-7%**		**-9%**	
Chi-Square	0.51		0.47		0.22		0.41	

Exhibit 2c: RF Passenger Fatalities Only, Side vs Frontal Comparisons, One to Four Years Before and After Beams Installed

Crash Type	One year before and after		Two years before and after		Three years before and after		Four years before and after	
	No Beam	Beam	No Beam	Beam	No Beam	Beam	No Beam	Beam
(1) SV NearSide	36	37	61	60	90	72	104	79
(2) SV FarSide	14	9	21	21	33	26	37	27
(3) SV Pure Frontal	77	79	139	133	203	155	225	165
(4) MV NearSide	47	57	104	97	158	119	171	132
(5) MV FarSide	34	24	62	36	86	50	94	52
(6) MV Pure Frontal	143	128	267	237	385	299	433	316
Total	351	334	654	584	955	721	1,064	771
All Side (1+2+4+5)	131	127	248	214	367	267	406	290
Frontal (3+6)	220	207	406	370	588	454	658	481
Effectiveness	-3%		5%		6%		2%	
Chi-Square	0.04		0.21		0.34		0.06	
Nearside (1+4)	83	94	165	157	248	191	275	211
Frontal (3+6)	220	207	406	370	588	454	658	481
Effectiveness	-20%		-4%		0%		-5%	
Chi-Square	1.07		0.11		0.00		0.20	
Far Side (2+5)	48	33	83	57	119	76	131	79
Frontal (3+6)	220	207	406	370	588	454	658	481
Effectiveness	27%		25%		17%		18%	
Chi-Square	1.64		2.31		1.42		1.55	
SV side (1+2)	50	46	82	81	123	98	141	106
SV Frontal (3)	77	79	139	133	203	155	225	165
Effectiveness	10%		-3%		-4%		-3%	
Chi-Square	0.18		0.03		0.06		0.02	

Exhibit 2c: RF Passenger Fatalities Only, Side vs Frontal Comparisons, One to Four Years Before and After Beams Installed continued

Crash Type	One year before and after		Two years before and after		Three years before and after		Four years before and after	
	No Beam	Beam	No Beam	Beam	No Beam	Beam	No Beam	Beam
SV Nearside (1)	36	37	61	60	90	72	104	79
SV Frontal (3)	77	79	139	133	203	155	225	165
Effectiveness	**0%**		**-3%**		**-5%**		**-4%**	
Chi-Square	0.00		0.02		0.06		0.04	
SV Far Side (2)	14	9	21	21	33	26	37	27
SV Frontal (3)	77	79	139	133	203	155	225	165
Effectiveness	**37%**		**-5%**		**-3%**		**0%**	
Chi-Square	1.06		0.02		0.01		0.00	
MV Side (4+5)	81	81	166	133	244	169	265	184
MV Frontal (6)	143	128	267	237	385	299	433	316
Effectiveness	**-12%**		**10%**		**11%**		**5%**	
Chi-Square	0.31		0.49		0.82		0.17	
MV Nearside (4)	47	57	104	97	158	119	171	132
MV Frontal (6)	143	128	267	237	385	299	433	316
Effectiveness	**-35%**		**-5%**		**3%**		**-6%**	
Chi-Square	1.73		0.09		0.05		0.17	
MV Far Side (5)	34	24	62	36	86	50	94	52
MV Frontal (6)	143	128	267	237	385	299	433	316
Effectiveness	**21%**		**35%**		**25%**		**24%**	
Chi-Square	0.66		3.50		2.24		2.18	

beams by the crash type. In each section, side data are presented with the corresponding counts of pure frontal crashes (single or multi-vehicle). The section labeled "All Side" includes both single and multi-vehicle side impacts. For this portion of the analysis, FARS calendar year data from 1989 through 2001 was used.

The data in Exhibit 2 are presented as:

Fatalities	No Beam	Beam
Side	N_{11}	N_{12}
Frontal (Control Group)	N_{21}	N_{22}

The ratio N_{22}/N_{21} is an indirect measure of the likelihood of post-standard light truck crashes relative to pre-standard vehicles. It takes the differences of exposure and the effects of any other safety-related improvements into account. The actual number of post-standard side impact fatalities is N_{12}. If the amended FMVSS 214 had no effect on side impacts, the expected number of fatalities in post-standard crashes would be $N_{11} \times (N_{22}/N_{21})$.

The change in side crashes relative to the frontals would be:

$$1 - \frac{N_{12}\Big/ N_{11}}{N_{22}\Big/ N_{21}}$$

The equation above is a measure of the effectiveness of the amended standard in side impacts, and is mathematically equivalent to the more easily computed:

$$\text{Effectiveness} = 1 - \frac{N_{12}N_{21}}{N_{22}N_{11}}$$

Note in Exhibits 2a and 2b that single vehicle side impacts consistently show the highest effectiveness levels. Specifically, single vehicle nearside impacts for drivers as well as all front outboard occupants combined show the highest effectiveness in each group of data (one to four years before and after), with point estimates (although not statistically significant) ranging from 7 to 21 percent. This makes intuitive sense, in that one would expect the beam to be best at protecting the occupant closest to it.

Fatalities in multi-vehicle crashes, as well as single vehicle far side impacts, show an inconsistent pattern across the number of years of included data. Driver fatalities actually increase in multi-vehicle side impacts, with point estimates ranging from a 2 to 14 percent increases, both near and far side. Data for right front passengers are inconsistent, most likely due to the small sample sizes involved. Far side impacts, particularly multi-vehicle far side, show the greatest improvements, ranging from 21 to 35 percent. None of these effectiveness estimates are statistically significant, however.

Results of chi-square tests are also shown for each comparison in Exhibit 2. With 1 degree of freedom (df) for each test, the chi-square value would need to be at least 3.841 to be statistically significant at the 0.05 level. While none of these comparisons is statistically significant, there is a consistent effect of improved performance for single vehicle side impacts for drivers and combined front outboard passengers, both nearside and far side and for each on the $\pm 1, \pm 2, \pm 3,$ and ± 4 comparisons. This is consistent with the results for passenger cars (Kahane, 1982).

The principal limitation of this analysis is that it had to be restricted to make-model year ranges when make-models were (1) never equipped with air bags or (2) always equipped with air bags. Many make-models could not be used at all (see Exhibit 1). Another limitation of this analysis is that the ratio of side to frontal impacts (and, thus, fatalities) may vary with vehicle age. If this is the case, the differences in the side vs. frontal ratios presented may in part be due to a vehicle age bias. In the section on regression analysis, vehicle age is one of the controlled variables.

Exposure Data

Although frontal fatalities appear to be a reasonable control group, it is possible that there could actually be some unexpected change in the control group of frontal impacts. Performing an additional analysis, without relying on comparisons to frontal impacts, would provide additional information. One method is to calculate the side impact fatality rate per one thousand (1,000) registered vehicle years, for a given set of vehicles in a given calendar year. Registration data through calendar year 2000 were available.

Since this analysis does not compare side impact fatalities to frontals, it seems at first glance that air bags are no longer a confounding factor and that it would be possible to use more data than the model year ranges shown in Exhibit 1. However, air bags can be effective in side impacts that have a frontal force component, and that especially includes many single-vehicle side impacts.

The average side impact fatality rate per 1,000 vehicle years then can be determined using FARS calendar years 1989 through 2000. Rates for each crash type were determined by dividing the appropriate number of fatalities by the vehicle exposure (number of registered vehicle years) divided by 1,000. The registration years from the Polk data are adjusted, for this analysis, by front-seat occupancy rates. Since the driver's seat is always occupied, and the right-front seat is occupied about one-third of the time, each vehicle registration year corresponds to one driver year, 0.33 right front passenger years, and 1.33 front-seat occupant years. Effectiveness was then determined by subtracting the rate for the "Beam" data from the rate for the "Before Beam" data, and dividing by the "Before Beam" rate. As in the previous section, model years 1990 through 1997 were used in this portion of the analysis. Also note that, because of varying times during the calendar year that a vehicle would be manufactured, counts were not used in which the model year was equal to the calendar year. This results in more standardized data across various manufacturers and different models.

The exposure data and fatality rates are presented in Exhibit 3a-c. FARS counts differ from those in Exhibit 2 in a number of ways. Since registration data were only available through model year 2000, vehicle data had to be limited through model year 2000 rather than 2001 as could be used in Exhibit 2. In addition, registration data cannot be separated as specifically as vehicle data that contains the VIN. Therefore, where similar vehicles (such as the various weights of Chevrolet C/K and GMC Sierra pickups) that had different usable model years of data could not be distinguished, the smallest set of usable model years had to be utilized to avoid including vehicles with modifications such as air bag installations. Finally, as mentioned previously, data in which the model year was equal to the calendar year were eliminated.

All front outboard occupants are shown in Exhibit 3a, drivers in 3b, and right front passengers in 3c. Recall that, because of differing years in which driver and right front passenger air bags were initially installed, the models years useable for drivers were not always the same as those for right front passengers. Because of this, vehicle exposure counts are not identical for drivers and right front passengers. In addition, since a driver is always present in a vehicle, but a passenger is not, the exposure counts must be adjusted in order for the rates for drivers and passengers to be comparable. Therefore, the counts for exposure were multiplied by one-third, to approximate the number of right front passengers typically found in relation to drivers. Data at the top of the exhibit are the numbers of occupant fatalities by type of crash; the vehicle exposure numbers are separated by pre- and post-standard vehicles ("No Beam" and "Beam"). As before, data are presented in four sets, from within one to four years before and after installation of side door beams.

Significance tests are provided in Exhibit 3 for the change in fatality rate after side door beams were installed. Since the fatality rate per one thousand registered vehicles is a Poisson distribution divided by a constant, the standard deviation is the square root of the number of fatalities divided by the constant. Significance is thus determined first by calculating the standard deviation of each population as:

$$s_i = \frac{\sqrt{N}}{R}$$

where s_i is the standard deviation of the distribution either before (i=b) or after (i=a) side door beams were installed, N is the number of fatalities in the crash type of interest (either before or after side beams), and R is the associated number of registered vehicles divided by a constant. Since the rates presented in Exhibit 3 are per one thousand vehicle miles, the constant is 1,000. The use of a constant does not influence the results of significance testing. The standard deviation of the difference between the two rates is then:

$$s = \sqrt{(s_b)^2 + (s_a)^2}$$

where s_b is the standard deviation (as calculated above) for the fatalities "Before" and s_a the standard deviation for fatalities "After." Significance is then tested as

$$z = \frac{r_b - r_a}{s}$$

Exhibit 3a: Front Outboard Occupants, Vehicle Exposure Crash Rates, 1989 through 2000

N of fatalities Crash Type	One year before and after		Two years before and after		Three years before and after		Four years before and after	
	No Beam	Beam	No Beam	Beam	No Beam	Beam	No Beam	Beam
SV Nearside	114	91	153	118	188	131	202	137
SV Farside	78	66	100	87	117	97	125	102
SV Frontal	343	321	477	415	558	484	578	498
MV Nearside	188	196	315	257	378	288	391	299
MV Farside	99	91	155	118	186	132	193	135
MV Frontal	600	505	867	679	1,009	785	1,060	807
Total	1,422	1,270	2,067	1,674	2,436	1,917	2,549	1,978
Vehicle Exposure**	21,863,333	21,356,667	33,756,667	29,370,000	40,413,333	32,940,000	42,916,667	33,896,667
All Frontal&Side Fatalities	1422	1,270	2067	1674	2436	1917	2549	1978
Rate	6.50%	5.95%	6.12%	5.70%	6.03%	5.82%	5.94%	5.84%
Effectiveness	9%		7%		3%		2%	
Significance test	*2.32		*2.19		1.15		0.59	
All Side Fatalities	479	444	723	580	869	648	911	673
Rate	2.19%	2.08%	2.14%	1.97%	2.15%	1.97%	2.12%	1.99%
Effectiveness	5%		8%		9%		6%	
Significance Test	0.80		1.46		*1.72		1.32	
Nearside Fatalities	302	287	468	375	566	419	593	436
Rate	1.38%	1.34%	1.39%	1.28%	1.40%	1.27%	1.38%	1.29%
Effectiveness	3%		8%		9%		7%	
Significance Test	0.33		1.19		1.50		1.14	
Far Side Fatalities	177	157	255	205	303	229	318	237
Rate	0.81%	0.74%	0.76%	0.70%	0.75%	0.70%	0.74%	0.70%
Effectiveness	9%		8%		7%		6%	
Significance Test	0.88		0.85		0.87		0.68	

*   statistically significant at one-tailed 0.05 level
** Vehicle registration years weight by front outboard seat occupancy (1.33)

Exhibit 3a continued: Front Outboard Occupants, Vehicle Exposure Crash Rates, 1989 through 2000

	One year before and after		Two years before and after		Three years before and after		Four years before and after	
SV side Fatalities	192	157	253	205	305	228	327	239
Rate	0.88%	0.74%	0.75%	0.70%	0.75%	0.69%	0.76%	0.71%
Effectiveness	**16%**		**7%**		**8%**		**7%**	
Significance Test	***1.66**		**0.76**		**0.99**		**0.92**	
SV Nearside Fatalities	114	91	153	118	188	131	202	137
Rate	0.52%	0.43%	0.45%	0.40%	0.47%	0.40%	0.47%	0.40%
Effectiveness	**18%**		**11%**		**15%**		**14%**	
Significance Test	**1.44**		**0.99**		**1.39**		**1.39**	
SV Far Side Fatalities	78	66	100	87	117	97	125	102
Rate	0.36%	0.31%	0.30%	0.30%	0.29%	0.29%	0.29%	0.30%
Effectiveness	**13%**		**0%**		**-2%**		**-3%**	
Significance Test	**0.86**		**0.00**		**-0.12**		**-0.24**	
MV Side Fatalities	287	287	470	375	564	420	584	434
Rate	1.31%	1.34%	1.39%	1.28%	1.40%	1.28%	1.36%	1.28%
Effectiveness	**-2%**		**8%**		**9%**		**6%**	
Significance Test	**-0.28**		**1.25**		**1.41**		**0.96**	
MV Nearside Fatalities	188	196	315	257	378	288	391	299
Rate	0.86%	0.92%	0.93%	0.88%	0.94%	0.87%	0.91%	0.88%
Effectiveness	**-7%**		**6%**		**7%**		**3%**	
Significance Test	**-0.64**		**0.77**		**0.87**		**0.42**	
MV Far Side Fatalities	99	91	155	118	186	132	193	135
Rate	0.45%	0.43%	0.46%	0.40%	0.46%	0.40%	0.45%	0.40%
Effectiveness	**6%**		**13%**		**13%**		**11%**	
Significance Test	**0.42**		**1.10**		**1.23**		**1.09**	

* statistically significant at one-tailed 0.05 level

Exhibit 3a continued: Front Outboard Occupants, Vehicle Exposure Crash Rates, 1989 through 2000

	One year before and after		Two years before and after		Three years before and after		Four years before and after	
All Frontal Crash Fatalities	943	826	1,344	1,094	1,567	1,269	1,638	1,305
Rate	4.31%	3.87%	3.98%	3.72%	3.88%	3.85%	3.82%	3.85%
Effectiveness	10%		6%		1%		-1%	
Significance Test	*2.29		1.64		0.17		-0.23	
SV Frontal Fatalities	343	321	477	415	558	484	578	498
Rate	1.57%	1.50%	1.41%	1.41%	1.38%	1.47%	1.35%	1.47%
Effectiveness	4%		0%		-6%		-9%	
Significance Test	0.55		0.00		-1.00		-1.42	
MV Frontal Fatalities	600	505	867	679	1,009	785	1,060	807
Rate	2.74%	2.36%	2.57%	2.31%	2.50%	2.38%	2.47%	2.38%
Effectiveness	14%		10%		5%		4%	
Significance Test	*2.47		*2.06		0.98		0.79	

* statistically significant at one-tailed 0.05 level

Exhibit 3b: Drivers, Vehicle Exposure Crash Rates, 1989 through 2000

N of fatalities Crash Type	One year before and after		Two years before and after		Three years before and after		Four years before and after	
	No Beam	Beam	No Beam	Beam	No Beam	Beam	No Beam	Beam
SV Nearside	81	56	104	69	116	76	119	76
SV Farside	64	57	80	69	87	75	91	79
SV Frontal	277	253	364	304	404	354	405	359
MV Nearside	149	143	228	172	246	184	247	187
MV Farside	68	69	99	86	108	92	109	93
MV Frontal	469	394	635	488	691	548	698	557
Total	1,108	972	1,510	1,188	1,652	1,329	1,669	1,351
Vehicle Exposure**	15,690,000	15,500,000	23,000,000	19,860,000	25,510,000	21,470,000	26,060,000	21,890,000
All Frontal&Side Fatalities	1108	972	1510	1188	1652	1329	1669	1351
Rate	7.06%	6.27%	6.57%	5.98%	6.48%	6.19%	6.40%	6.17%
Effectiveness	11%		9%		4%		4%	
Significance Test	*2.71		*2.41		1.23		1.01	
All Side Fatalities	362	325	511	396	557	427	566	435
Rate	2.31%	2.10%	2.22%	1.99%	2.18%	1.99%	2.17%	1.99%
Effectiveness	9%		10%		9%		9%	
Significance Test	1.25		1.62		1.46		1.40	
Nearside Fatalities	230	199	332	241	362	260	366	263
Rate	1.47%	1.28%	1.44%	1.21%	1.42%	1.21%	1.40%	1.20%
Effectiveness	12%		16%		15%		14%	
Significance Test	1.37		*2.07		*1.97		*1.95	
Far Side Fatalities	132	126	179	155	195	167	200	172
Rate	0.84%	0.81%	0.78%	0.78%	0.76%	0.78%	0.77%	0.78%
Effectiveness	3%		0%		-2%		-2%	
Significance Test	0.28		-0.03		-0.16		-0.23	

* statistically significant at one-tailed 0.05 level

** Vehicle registration years

Exhibit 3b continued: Drivers, Vehicle Exposure Crash Rates, 1989 through 2000

	One year before and after		Two years before and after		Three years before and after		Four years before and after	
SV side Fatalities	145	113	184	138	203	151	210	155
Rate	0.92%	0.73%	0.80%	0.69%	0.80%	0.70%	0.81%	0.71%
Effectiveness	**21%**		**13%**		**12%**		**12%**	
Significance Test	***1.90**		**1.26**		**1.16**		**1.23**	
SV Nearside Fatalities	81	56	104	69	116	76	119	76
Rate	0.52%	0.36%	0.45%	0.35%	0.45%	0.35%	0.46%	0.35%
Effectiveness	**30%**		**23%**		**22%**		**24%**	
Significance Test	***2.07**		***1.72**		***1.72**		***1.89**	
SV Far Side Fatalities	64	57	80	69	87	75	91	79
Rate	0.41%	0.37%	0.35%	0.35%	0.34%	0.35%	0.35%	0.36%
Effectiveness	**10%**		**0%**		**-2%**		**-3%**	
Significance Test	**0.57**		**0.01**		**-0.15**		**-0.21**	
MV Side Fatalities	217	212	327	258	354	276	356	280
Rate	1.38%	1.37%	1.42%	1.30%	1.39%	1.29%	1.37%	1.28%
Effectiveness	**1%**		**9%**		**7%**		**6%**	
Significance Test	**0.12**		**1.09**		**0.96**		**0.83**	
MV Nearside Fatalities	149	143	228	172	246	184	247	187
Rate	0.95%	0.92%	0.99%	0.87%	0.96%	0.86%	0.95%	0.85%
Effectiveness	**3%**		**13%**		**11%**		**10%**	
Significance Test	**0.25**		**1.34**		**1.22**		**1.08**	
MV Far Side Fatalities	68	69	99	86	108	92	109	93
Rate	0.43%	0.45%	0.43%	0.43%	0.42%	0.43%	0.42%	0.42%
Effectiveness	**-3%**		**-1%**		**-1%**		**-2%**	
Significance Test	**-0.16**		**-0.04**		**-0.09**		**-0.11**	

* statistically significant at one-tailed 0.05 level

Exhibit 3b continued: Drivers, Vehicle Exposure Crash Rates, 1989 through 2000

	One year before and after		Two years before and after		Three years before and after		Four years before and after	
All Frontal Crash Fatalities	746	647	999	792	1,095	902	1,103	916
Rate	4.75%	4.17%	4.34%	3.99%	4.29%	4.20%	4.23%	4.18%
Effectiveness	12%		8%		2%		1%	
Significance Test	*2.43		*1.80		0.48		0.26	
SV Frontal Fatalities	277	253	364	304	404	354	405	359
Rate	1.77%	1.63%	1.58%	1.53%	1.58%	1.65%	1.55%	1.64%
Effectiveness	8%		3%		-4%		-6%	
Significance Test	0.90		0.43		-0.55		-0.74	
MV Frontal Fatalities	469	394	635	488	691	548	698	557
Rate	2.99%	2.54%	2.76%	2.46%	2.71%	2.55%	2.68%	2.54%
Effectiveness	15%		11%		6%		5%	
Significance Test	*2.38		*1.94		1.04		0.90	

* statistically significant at one-tailed 0.05 level

Exhibit 3c: RF Passenger Fatalities, Vehicle Exposure Crash Rates, 1989 through 2000

N of fatalities Crash Type	One year before and after		Two years before and after		Three years before and after		Four years before and after	
	No Beam	Beam	No Beam	Beam	No Beam	Beam	No Beam	Beam
SV Nearside	33	35	49	49	72	55	83	61
SV Farside	14	9	20	18	30	22	34	23
SV Frontal	66	68	113	111	154	130	173	139
MV Nearside	39	53	87	85	132	104	144	112
MV Farside	31	22	56	32	78	40	84	42
MV Frontal	131	111	232	191	318	237	362	250
Total	314	298	557	486	784	588	880	627
Vehicle Exposure**	6,173,333	5,856,667	10,756,667	9,510,000	14,903,333	11,470,000	16,856,667	12,006,667
All Frontal&Side Fatalities	314	298	557	486	784	588	880	627
Rate	5.09%	5.09%	5.18%	5.11%	5.26%	5.13%	5.22%	5.22%
Effectiveness	0%		1%		3%		0%	
Significance Test	0.00		0.21		0.47		-0.01	
All Side Fatalities	117	119	212	184	312	221	345	238
Rate	1.90%	2.03%	1.97%	1.93%	2.09%	1.93%	2.05%	1.98%
Effectiveness	-7%		2%		8%		3%	
Significance Test	-0.53		0.18		0.95		0.38	
Nearside Fatalities	72	88	136	134	204	159	227	173
Rate	1.17%	1.50%	1.26%	1.41%	1.37%	1.39%	1.35%	1.44%
Effectiveness	-29%		-11%		-1%		-7%	
Significance Test	-1.59		-0.89		-0.12		-0.67	
Far Side Fatalities	45	31	76	50	108	62	118	65
Rate	0.73%	0.53%	0.71%	0.53%	0.72%	0.54%	0.70%	0.54%
Effectiveness	27%		26%		25%		23%	
Significance Test	1.38		1.64		*1.88		*1.70	

* statistically significant at one-tailed 0.05 level

** Vehicle registration years weighted by right-front seat occupancy (0.33)

Exhibit 3c continued: RF Passenger Fatalities, Vehicle Exposure Crash Rates, 1989 through 2000

	One year before and after		Two years before and after		Three years before and after		Four years before and after	
SV side Fatalities	47	44	69	67	102	77	117	84
Rate	0.76%	0.75%	0.64%	0.70%	0.68%	0.67%	0.69%	0.70%
Effectiveness	1%		-10%		2%		-1%	
Significance Test	0.06		-0.55		0.13		-0.06	
SV Nearside Fatalities	33	35	49	49	72	55	83	61
Rate	0.53%	0.60%	0.46%	0.52%	0.48%	0.48%	0.49%	0.51%
Effectiveness	-12%		-13%		1%		-3%	
Significance Test	-0.46		-0.61		0.04		-0.19	
SV Far Side Fatalities	14	9	20	18	30	22	34	23
Rate	0.23%	0.15%	0.19%	0.19%	0.20%	0.19%	0.20%	0.19%
Effectiveness	32%		-2%		5%		5%	
Significance Test	0.92		-0.05		0.17		0.19	
MV Side Fatalities	70	75	143	117	210	144	228	154
Rate	1.13%	1.28%	1.33%	1.23%	1.41%	1.26%	1.35%	1.28%
Effectiveness	-13%		7%		11%		5%	
Significance Test	-0.73		0.62		1.08		0.51	
MV Nearside Fatalities	39	53	87	85	132	104	144	112
Rate	0.63%	0.90%	0.81%	0.89%	0.89%	0.91%	0.85%	0.93%
Effectiveness	-43%		-11%		-2%		-9%	
Significance Test	-1.70		-0.65		-0.18		-0.69	
MV Far Side Fatalities	31	22	56	32	78	40	84	42
Rate	0.50%	0.38%	0.52%	0.52%	0.52%	0.35%	0.50%	0.35%
Effectiveness	25%		35%		33%		30%	
Significance Test	1.05		*2.01		*2.16		*1.94	

* statistically significant at one-tailed 0.05 level

Exhibit 3c continued: RF Passenger Fatalities, Vehicle Exposure Crash Rates, 1989 through 2000

	One year before and after		Two years before and after		Three years before and after		Four years before and after	
All Frontal Crash Fatalities	197	179	345	302	472	367	535	389
Rate	3.19%	3.06%	3.21%	3.18%	3.17%	3.20%	3.17%	3.24%
Effectiveness	4%		1%		-1%		-2%	
Significance Test	0.42		0.13		-0.15		-0.31	
SV Frontal Fatalities	66	68	113	111	154	130	173	139
Rate	1.07%	1.16%	1.05%	1.17%	1.03%	1.13%	1.03%	1.16%
Effectiveness	-9%		-11%		-10%		-13%	
Significance Test	-0.48		-0.79		-0.77		-1.05	
MV Frontal Fatalities	131	111	232	191	318	237	362	250
Rate	2.12%	1.90%	2.16%	2.01%	2.13%	2.07%	2.15%	2.08%
Effectiveness	11%		7%		3%		3%	
Significance Test	0.88		0.73		0.38		0.38	

* statistically significant at one-tailed 0.05 level

where r_b is the fatality rate before side beams, and r_a the rate after. Values above 1.65 are considered statistically significant at the 0.05 level in these one-tailed tests.

The majority of front outboard occupants in Exhibit 3a are drivers, and thus the data are similar to Exhibit 3b, which includes drivers only. Results in both of these exhibits show statistically significant decreases in fatality rates for "All crashes" after side door beams are installed, within one to two years of beam installation, ranging from 7 to 11 percent. Single vehicle side impact fatalities were reduced a statistically significant 16 percent for front outboard occupants within one year of beam installation. Single vehicle nearside impact fatalities were reduced 11 to 18 percent and, while not statistically significant, were generally the largest reductions for each group of front outboard occupants. In addition, the "all" frontal as well as multi-vehicle frontal fatality rates had a statistically significantly decrease within one to two years of adding the side beams, for front outboard occupants as well as drivers alone. The results in frontal impacts likely have more to do with the installation of anti-lock braking systems than to side door beams.

For drivers only, all four years of single vehicle nearside fatality rates had statistically significant decreases after beams were installed, with reductions ranging from 22 to 30 percent. These are generally higher than any other effects shown in Exhibit 3b. Reductions for all nearside fatalities (which includes both multi- and single vehicle) ranged from 12 to 16 percent. Comparisons using data two or more years beyond the installation of the beams were statistically significant. In contrast, changes in single vehicle frontal fatality rates never exceeded ± 8 percent.

Right front passengers seem to demonstrate a different pattern, as shown in Exhibit 3c. Multi-vehicle far side fatality rates showed statistically significant decreases when more than one year of data for each group was used, ranging from 30 to 35 percent. Within one year of beam installation, multi-vehicle far side fatalities decreased a non-statistically significant 25 percent. In addition, the "all" far side fatality rates decreased from 23 to 27 percent for right front passengers, statistically significant when three or four years of data were used. Significance tests are dependent on sample size, which clearly is influencing the results for these small samples of right front passengers.

The vehicles without side beams are, in general, older vehicles, and tend to be overrepresented in crashes. "All" and single vehicle frontal crashes also show improvement for front outboard occupants, particularly for drivers. Although not statistically significant, multi-vehicle frontal fatalities were reduced for right front passengers after beam installation. This is important as fatalities in frontal impact were used as the control in the previous analysis. Any rate comparisons would need to take these tendencies into account. The following regression analysis enables this to be accomplished.

Regression Analysis

A logistic regression analysis would enable the full range of model years to be used in analyzing the ratio of side to frontal crash fatalities. In the previous analyses, if the number of model years of usable data was restricted for either the "Before" or "After" group because of air bag or production limitations, the other group was likewise restricted, in order to keep the counts and rates consistent. Using regression, vehicles can simply be coded as "Before" or "After" without restricting the dataset. Vehicle age can be entered as an additional variable in order to control for possible bias. Thus, regression allows the data to be analyzed without the age bias as well as increase the number of vehicles in the dataset. Most important, "Air bag" can be included as a control variable. This permits adding make-models that were equipped with side beams and air bags simultaneously, and using the full range of model years even if air bags were first installed during those years.

The preceding analyses of the effectiveness of side door beams may have contained a bias against the amended standard, since the post-standard vehicles are newer than those manufactured before the standard took effect. Newer vehicles have a higher ratio of side to frontal impacts than do older vehicles (Kahane, 1982, p 253). Thus, a higher ratio of side to frontal crashes would be expected in the post-standard trucks without any changes in the vehicles themselves. This could possibly mask the effect of the side door beams, working against their possible reduction in side impact fatalities.

Therefore, it is important to control for the vehicle age effect. Because of year-to-year changes in roads and demographics, calendar year is also another potential variable that could influence analysis results. However, these two variables (vehicle age and calendar year) are clearly related to one another, which could cause a problem if both are used in a regression analysis. As calendar year increases, obviously vehicle age also increases for the same model year. Since they vary together, what is actually related to one of the variables could possibly be attributed to the other one. This co-linearity is a potential problem for the analysis.

Looking at the effect each of these two variables actually has on fatality crash data will help determine each one's relative importance in the analysis. Because it is an attribute of crashes that is being determined, and not anything specifically related to the installation of side door beams, it is unnecessary to limit the data to specific makes or model years. In fact, for a larger and more descriptive sample, passenger cars can be included. Exhibit 4 presents passenger vehicle fatalities in single vehicle crashes (nearside and pure frontal impacts) by calendar year. Passenger vehicles include passenger cars as well as light trucks, vans, and SUVs. An effect value is calculated for each calendar year relative to the year 1997, which serves as the comparison year in the regression analyses to follow. The effect value is determined as:

$$\text{Effect} = \ln\left(\frac{N_{S_i}/N_{F_i}}{N_{S_c}/N_{F_c}}\right)$$

where
N_{Si} is the number of fatalities in nearside impacts in year i
N_{Fi} is the number of fatalities in frontal impacts in year i
N_{Sc} is the number of fatalities in nearside impacts in 1997
N_{Fc} is the number of fatalities in frontal impacts in 1997

Although the present study concerns only light trucks, passenger cars are included in the data for this exhibit to increase the sample size, in order to get a more representative view of any possible calendar year effect. In addition, a wider span of model years is also included to increase the sample size.

Exhibit 4: Model Year 1985 through 2001 Passenger Vehicle Fatalities

	SV Nearside	SV Pure Frontal	Effect
1989	664	1,515	0.122
1990	757	1,734	0.118
1991	861	1,837	0.189
1992	828	1,999	0.065
1993	912	2,171	0.079
1994	995	2,293	0.112
1995	1,164	2,706	0.103
1996	1,291	2,874	0.146
1997	1,200	3,092	0.000
1998	1,270	3,134	0.043
1999	1,395	3,106	0.146
2000	1,402	3,391	0.063
2001	1,409	3,489	0.040

Note that the effect column shows neither a strong effect nor a trend. The ratio of single vehicle nearside impacts relative to frontals does not consistently increase, nor is there a particularly strong effect in any given year or set of years.

Exhibit 5 presents the effect information as calculated above for additional vehicle and model year combinations. Data from Exhibit 4 are repeated for ease of comparison, in the final column. Exhibit 5 presents columns for light trucks alone as well as for all passenger vehicles. In addition, data for the limited set of model years that will be usable for the regression analysis (1989 through 1997) are presented in addition to the wider span of model years used above. For all data sets, 1997 serves as the comparison year. Note that, regardless of the set of vehicles or model years included, there is no trend to the data, nor is there any particular calendar year or years that stands out as being particularly strong. The possible exceptions might be the earlier few calendar years where light trucks only are included. The data for these cells, however, is rather sparse. For example, for calendar year 1989, there is a total of 121 fatalities when model year 1989 through 2000 are included, 37 in single vehicle nearside crashes and 84 in frontal. In comparison, there are over 1,300 fatalities for calendar year 2001 in the same dataset. Although there are higher effects noted for some of these earlier years, it is likely this is a spurious result of the small amount of data available.

Similar calculations can be performed for the vehicle age factor. Exhibit 6 presents, for the same data used above, effect values by vehicle age. Four-year-old vehicles were arbitrarily chosen as the reference set, to be somewhat consistent with the tables above. Note, however, that for either set of model years used and either set of vehicles, there is a strong, persistent vehicle age effect. Exhibit 6 clearly shows that, as vehicles age, there is a much higher proportion of fatalities in frontal impacts relative to side impacts, shown by the higher, negative entries in the effect column for older vehicles. The positive values for the newer vehicles shows that these vehicles have a relatively higher number of side impact fatalities. There is a clear negative relationship between the proportion of side impact fatalities and vehicle age.

Exhibit 5: Calendar Year Effect for Fatalities by Model Year

Calendar Year	LTV only MY 1989-1997	Passenger Vehicles MY 1989-1997	LTV only MY 1985-2001	Passenger Vehicles MY 1985-2001
1989	0.440	0.192	0.325	0.122
1990	0.177	0.050	0.294	0.118
1991	0.320	0.240	0.289	0.189
1992	0.048	0.117	0.090	0.065
1993	0.058	0.067	0.161	0.079
1994	0.226	0.076	0.174	0.112
1995	0.047	0.118	0.129	0.103
1996	0.187	0.164	0.180	0.146
1997	0.000	-0.003	0.000	0.000
1998	0.121	0.047	0.095	0.043
1999	0.154	0.128	0.183	0.146
2000	-0.068	-0.004	0.083	0.063
2001	-0.025	-0.018	0.042	0.400

Exhibit 6: Vehicle Age Effect for Fatalities by Model Year

Vehicle Age	LTV only MY 1989-1997	Passenger Vehicles MY 1989-1997	LTV only MY 1985-2001	Passenger Vehicles MY 1985-2001
16	-	-	-0.244	-0.487
15	-	-	-0.296	-0.270
14	-	-	-0.173	-0.356
13	-	-	-0.275	-0.292
12	-0.327	-0.482	-0.453	-0.399
11	-0.211	-0.318	-0.094	-0.232
10	-0.243	-0.303	-0.255	-0.185
9	-0.213	-0.254	-0.097	-0.190
8	-0.118	-0.199	-0.152	-0.148
7	-0.174	-0.173	-0.107	-0.133
6	-0.074	-0.156	-0.113	-0.114
5	0.018	-0.079	0.062	-0.016
4	0.000	0.000	0.000	0.000
3	0.151	0.104	0.054	0.087
2	0.146	0.066	0.099	0.080
1	0.153	0.042	0.169	0.097
0	0.216	0.199	0.182	0.225

The effect columns in Exhibits 5 and 6 show what one would expect as the maximum likelihood estimate in a logistic regression. Given that there may be colinearity issues between the calendar year and vehicle age variables, and knowing that there is a true and strong effect of vehicle age on the data, a logistic regression is first presented without including calendar year in the model. While it would be advantageous to be able to control for this as well, the potential problems that might arise could outweigh the benefits of its inclusion.

The dependent variable in this regression analysis is the damage location, coded to represent either side damage or frontal damage. If side beams were effective in preventing side impact fatalities, then one would expect relatively fewer side impact fatalities in relation to frontal impact fatalities. The probability modeled is that the damage location is side and not frontal. In addition, the variable in the model representing the presence of the beams should be statistically significant, indicating that the beams play a definitive role in the reduction of fatalities in side impacts relative to those in frontals. The independent variables used were:

BEAM = 1 if vehicle was equipped with side door beam, 0 otherwise

AB = 1 if air bag equipped in that vehicle at that seating position, 0 otherwise

ABS = 1 if vehicle was equipped with four-wheel antilock brake system, 0 otherwise

RWAL = 1 if vehicle was equipped with rear-wheel antilock brake system, 0 otherwise

RFPASS = 1 if the seating position was right front passenger, 0 if driver

FEMALE = 1 if the occupant was female, 0 if male

AGE = age of the occupant

AGE_SQD to account for possible nonlinearity of the age effect

VEHAGE = age of the vehicle

BRANDNEW = 1 if vehicle was less than a year old at the time of the crash, 0 otherwise

BELT = 1 if occupant was restrained with a safety belt, 0 otherwise

CURBWT = curb weight of truck

SUV = 1 if the vehicle was a sport utility vehicle, 0 otherwise

VAN = 1 if the vehicle was a van, 0 otherwise

FOURWD = 1 if vehicle has four-wheel (or all-wheel) drive, 0 otherwise

Note that there is no separate variable for pickup trucks. Vans and utility vehicles are coded as above, while pickups would have a value of '0' on both variables. Thus, all vehicle types are defined.

Exhibit 7 presents parameter estimates for the logistic regression for single vehicle nearside impact fatalities relative to single vehicle frontal fatalities, using the set of independent variables above. This is the type of crash most likely to benefit from the presence of the side door beams. Data from calendar years 1989 through 2001 were included. Again, the regression procedure allows a larger sample size than the previous methodologies, since data before and after beam installation do not have to be matched for such situations as presence of air bags or years of production. The number of observations in the single vehicle nearside regression, 3,559 (2,760 frontal and 799 side impacts), is much larger than the 1687 observations (1,286 frontal and 401 side) for four years before and after in Exhibit 2a, or the 339 nearside fatalities used to compute the four-year exposure rates in exhibit 3a. The overall chi-square for the model for all crashes was 151.5173, with a probability below 0.0001. Results of chi-square tests are also shown for each comparison in Exhibit 7. With 1 degree of freedom (df) for each test, the chi-square value would need to be at least 3.841 to be statistically significant at the 0.05 level. Parameter estimates with a chi-square probability less than 0.05 are set in bold in Exhibit 7. This shows that the presence of side door beams is associated with statistically significantly decreases in the proportion of single-vehicle nearside fatalities. In addition, note that the vehicle age is statistically significant, which would be expected given the data previously presented.

Exhibit 7: Maximum Likelihood Estimates, Logistic Regression without Calendar Year Variables

Parameter	SV Nearside	Chi-Square
Intercept	**-1.4273**	**17.57**
BEAM	**-0.2901**	**7.27**
AB	**0.3440**	**6.83**
ABS	0.1150	0.35
RWAL	-0.2237	2.02
RFPASS	**0.6453**	**34.61**
FEMALE	0.1388	1.56
AGE	0.0167	1.99
AGE_SQD	**-0.0004**	**8.39**
VEHAGE	**-0.0500**	**7.59**
BRANDNEW	0.0142	0.01
BELT	**0.4152**	**18.97**
CURBWT	0.0001	1.82
SUV	-0.2348	2.15
VAN	**-0.3852**	**6.58**
FOURWD	-0.0675	0.37

Bold entries are statistically significant at 2-tailed chi-square, α=0.05 level.

The reduction in fatalities due to effectiveness, or beam installation, can be calculated as

$$\text{Effectiveness} = 1 - e^b$$

where b is the estimate for beam in the regression equation. These values are also reported as the odds ratio estimate in the regression model. For single vehicle nearside impacts, this would be

$$\text{Effectiveness} = 1 - e^{-0.2901} = 0.2518$$

Thus, the effectiveness of beams in single vehicle nearside impacts is estimated to be 25 percent. That is, with side door beams you have a 25 percent lower chance of fatality as compared to a pre-standard light truck, in a single vehicle nearside impact, using frontal crashes as controls for both groups.

The vehicle age effect of –0.05 is statistically significant, consistent with the data shown in Exhibit 6. There is a definite trend for the proportion of side impacts to decrease (relative to frontal impacts) as vehicle age increases. The fact that both the vehicle age and beam estimates in Exhibit 7 are negative reflects the fact that, for both factors, side impacts decrease relative to frontal impacts due to the effect of each factor.

Exhibit 8 presents the data for various crash types (all crashes combined, as well as separated by single or multi-vehicle, near or far-side crash) including single vehicle nearside crashes as has been shown above in Exhibit 7, for drivers and right front passengers combined. In addition, the estimated effectiveness of the beam (as calculated above) and the Chi-square value (Beam X^2) for the beam estimate are reported. In the final rows of each table, standard error (Beam s) and the sample size for side impact fatalities (N side imp fatalities) are presented.

Since earlier analyses showed differences between drivers and right front passengers, analyses will also be presented separately by seat position. Exhibit 9 shows the same model and crash types for drivers only, and Exhibit 10 for right front passengers only. Exhibits 9 and 10 have no RFPASS variable, since within those analyses observations do not vary by seat position. The statistically significant value of 0.6453 for RFPASS in Exhibit 8 reflects the fact that crash type prevalence varies by seating position. Recall that the dependent variable for the regression model is the proportion of side to frontal fatalities. Since the estimate is coded as being a passenger, and has a positive value, the interpretation is that the proportion of single vehicle nearside impact fatalities to frontal fatalities for right front passengers is greater than it is for drivers. Note that this has no bearing on the effect of the beam by seat position.

The regression analysis shows that beams are effective for drivers alone and in combination with right front passengers in single vehicle side impacts, particularly those on the nearside. The fatality reduction in single vehicle crashes attributed to the beams was 19 percent for both drivers alone and for all front outboard occupants. The reduction in fatalities credited to beams in single vehicle nearside impacts was 26 percent for drivers alone (Exhibit 9) and 25 percent for all front outboard passengers, as compared to single vehicle frontal impacts. All four of these effectiveness estimates are statistically significant. Although not statistically significant, the reduction for right front passengers alone was 18 percent for all single vehicle crashes, and 17 percent for single vehicle nearside crashes, certainly consistent with the other regression findings. These data show that equipping light trucks with side door beams has been effective in preventing fatalities in single vehicle nearside crashes, but has had little effect in multi-vehicle

Exhibit 8: Maximum Likelihood Estimates by Crash Type, Drivers and Right Front Passengers

Parameter	All Side	SV Side	MV Side	SV Nearside	MV Nearside	SV Farside	MV Farside
Intercept	0.0308	**-0.7368**	0.2196	**-1.4273**	-0.4224	**-1.5276**	-0.4500
BEAM	-0.0915	**-0.2150**	0.0163	**-0.2901**	0.0210	-0.1163	-0.0073
AB	**0.3027**	**0.1643**	**0.3455**	**0.3440**	**0.2865**	-0.0820	**0.4808**
ABS	-0.0090	0.2609	-0.1477	0.1150	-0.2504	**0.4929**	0.0712
RWAL	-0.1080	-0.1694	-0.0457	-0.2237	-0.1542	-0.0867	0.1933
RFPASS	**0.1929**	**0.2590**	0.1360	**0.6453**	**0.1680**	**-0.3794**	0.0637
FEMALE	**0.3562**	**0.2072**	**0.4337**	0.1388	**0.4212**	**0.2836**	**0.4596**
AGE	**-0.0223**	0.0107	**-0.0273**	0.0167	**-0.0265**	0.0069	**-0.0273**
AGE_SQD	**0.0002**	-0.0003	**0.0004**	-0.0004	**0.0004**	-0.0002	**0.0004**
VEHAGE	**-0.0314**	**-0.0356**	**-0.0290**	**-0.0500**	-0.0250	-0.0186	**-0.0358**
BRANDNEW	-0.0052	-0.0216	0.0134	0.0142	0.0974	-0.1150	-0.1526
BELT	0.0118	0.0601	-0.0221	**0.4152**	0.1175	**-0.5311**	**-0.3234**
CURBWT	-0.0001	0.0001	**-0.0002**	0.0001	-0.0001	0.0001	**-0.0003**
SUV	0.0636	**-0.3378**	**0.3709**	-0.2348	**0.4239**	**-0.5144**	0.2503
VAN	**-0.2126**	**-0.4194**	-0.0912	**-0.3852**	-0.0900	**-0.5097**	-0.1134
FOURWD	0.0188	-0.0143	0.0168	-0.0675	-0.0013	0.0477	0.0415
Effectiveness	0.09	**0.19**	-0.02	**0.25**	-0.02	0.11	0.01
Beam X^2	3.03	**6.01**	0.06	**7.27**	0.08	0.92	0.01
Beam s	0.0526	0.0877	0.0667	0.1076	0.0765	0.1215	0.1015
N side imp fatalities	3,633	1,376	2,257	799	1,495	577	762

Exhibit 9: Maximum Likelihood Estimates by Crash Type, Drivers Only

Parameter	All Side	SV Side	MV Side	SV Nearside	MV Nearside	SV Farside	MV Farside
Intercept	-0.0839	**-0.8425**	-0.0128	**-1.6272**	**-0.8006**	**-1.5071**	-0.4241
BEAM	-0.0765	**-0.2070**	0.0484	**-0.3056**	0.0251	-0.0901	0.0895
AB	**0.3141**	0.1392	**0.3987**	**0.3357**	**0.4043**	-0.1255	**0.4060**
ABS	-0.0513	0.2569	-0.2455	0.0478	**-0.4785**	**0.5360**	0.1742
RWAL	-0.1549	-0.2187	-0.0898	-0.2812	-0.2369	-0.1415	0.1971
FEMALE	**0.4030**	0.2129	**0.5020**	0.1531	**0.5370**	0.2361	**0.4260**
AGE	**-0.0276**	0.0132	**-0.0312**	0.0200	**-0.0297**	0.0085	**-0.0333**
AGE_SQD	**0.0003**	-0.0003	**0.0004**	-0.0004	**0.0004**	-0.0002	**0.0004**
VEHAGE	**-0.0287**	**-0.0328**	**-0.0261**	**-0.0581**	-0.0266	-0.0042	-0.0253
BRANDNEW	-0.0641	-0.0762	-0.0433	-0.0842	0.0135	-0.0949	-0.1616
BELT	0.0004	0.0407	-0.0417	**0.4504**	0.1128	**-0.5888**	**-0.3620**
CURBWT	0.0001	0.0001	-0.0001	0.0002	0.0000	0.0001	**-0.0003**
SUV	0.0759	**-0.2950**	**0.3794**	-0.1362	**0.5244**	**-0.5027**	0.0950
VAN	**-0.2779**	**-0.3525**	**-0.2331**	-0.2699	**-0.3397**	**-0.4859**	-0.0494
FOURWD	-0.0293	-0.0337	-0.0502	-0.1067	-0.0902	0.0493	0.0056
Effectiveness	0.07	**0.19**	-0.05	**0.26**	-0.03	0.09	-0.09
Beam X^2	1.57	**4.21**	0.38	**5.65**	0.08	0.46	0.57
Beam s	0.0611	0.1009	0.0780	0.1285	0.0903	0.1334	0.1183
N side imp fatalities	2,789	1,096	1,693	590	1,107	506	586

Bold entries are statistically significant at 2-tailed chi-square, $\alpha=0.05$ level

Exhibit 10: Maximum Likelihood Estimates by Crash Type,
Right Front Passengers Only

Parameter	All Side	SV Side	MV Side	SV Nearside	MV Nearside	SV Farside	MV Farside
Intercept	**0.6979**	-0.1114	**0.9949**	-0.2403	0.7278	**-2.1715**	-0.3849
BEAM	-0.1065	-0.1969	-0.0330	-0.1871	0.0776	-0.2799	-0.2319
AB	-0.0254	-0.0963	-0.1243	-0.1202	-0.3953	0.1933	0.4378
ABS	0.0455	0.3405	-0.0350	0.3639	0.0807	0.3969	-0.3781
RWAL	-0.0041	0.0589	0.0254	-0.0306	0.0049	0.4034	0.0827
FEMALE	**0.3441**	0.1969	**0.4406**	0.1384	**0.3833**	0.3745	**0.5809**
AGE	-0.0078	-0.0003	-0.0105	0.0022	-0.0077	0.0015	-0.0160
AGE_SQD	0.0000	-0.0002	0.0001	-0.0003	0.0000	-0.0001	0.0002
VEHAGE	**-0.0435**	-0.0495	-0.0401	-0.0235	-0.0219	**-0.1315**	**-0.0742**
BRANDNEW	0.1572	0.2243	0.1573	0.3777	0.2977	-0.2843	-0.1700
BELT	0.0800	0.2290	0.0493	0.3780	0.1311	-0.1453	-0.1756
CURBWT	**-0.0003**	0.0000	**-0.0004**	0.0000	-0.0005	0.0002	-0.0003
SUV	0.0432	-0.5687	0.4230	-0.5785	**0.2960**	-0.6602	**0.7171**
VAN	0.0253	**-0.8001**	**0.3074**	**-0.8571**	0.4904	-0.6942	-0.1106
FOURWD	0.1981	0.0979	0.2521	0.0807	**0.3132**	0.1685	0.1644
Effectiveness	0.10	0.18	0.03	0.17	-0.08	0.24	0.21
Beam X^2	0.99	1.10	0.06	0.81	0.26	0.84	1.32
Beam s	0.1070	0.1879	0.1334	0.2086	0.1516	0.3049	0.2015
N side imp fatalities	844	280	564	209	388	71	176

Bold entries are statistically significant at 2-tailed chi-square, $\alpha=0.05$ level.

crashes. There may possibly be an effect in single vehicle far side crashes – none of the results were statistically significant, but they consistently show positive results, particularly for right front passengers (for whom the impact would be on the driver's side).

The significance of the Van and SUV variables is of some interest. Remember that, for each of these variables, the type of vehicle is compared to pickup trucks. The statistically significant value of –0.3852 for VAN in Exhibit 8 denotes that Vans experience a smaller proportion (since the coefficient is negative) of single vehicle nearside impact fatalities relative to frontal fatalities than do pickup trucks. Looking at all side impacts, both vans and SUV's have a statistically significant negative coefficient, showing they have relatively fewer side impacts than frontals, when compared to pickup trucks. Again, this has no reflection on the effectiveness of the beam in different vehicle types, only the pattern of crashes resulting in fatalities.

In order to shed some light on whether the beams vary in effectiveness by type of vehicle, separate regression analyses were run for vans, pickup trucks, and SUVs, for single vehicle side impacts as well as single vehicle nearside impacts. The beam variable was statistically significant only for the group of pickup trucks, most likely a function of sample size, since there were far more pickup trucks in the sample than vans or SUVs. The number of single vehicle side impact fatalities in pickups was 1059, while there were only 177 in SUVs and 140 in vans.

Also of interest to note in the regression analysis is that, for drivers alone or in combination with right front passengers, the air bag variable (AB) was statistically significant in most of the crash types, particularly multi-vehicle crashes, while the variables representing antilock brake system (ABS and RWAL) were statistically significant in only a small number of crash configurations and less consistently. This increases confidence in previous analyses, in which changes in air bag configurations were accounted for but antilock brakes were not. Seating position, gender, restraint use and occupant age appear to be important factors to take into account.

Exhibit 11 presents an alternative model on the same set of data, but includes a set of dummy variables to control for calendar year. As in Exhibit 4, 1997 is used as the comparison year. Thus, the variable named CY1989 represents the effect of calendar year 1989, relative to 1997. The same is true for the remaining calendar year variables. Since each year is compared to 1997, no variable is needed for the

Exhibit 11: Maximum Likelihood Estimates, Logistic Regression with Calendar Year Variables

Parameter	SV Nearside
Intercept	**-2.1040**
BEAM	-0.0869
AB	**0.4042**
ABS	0.2143
RWAL	-0.2175
RFPASS	**0.6559**
FEMALE	0.1341
AGE	0.0162
AGE_SQD	**-0.0004**
VEHAGE	0.0242
BRANDNEW	-0.0071
BELT	**0.4339**
CURBWT	0.0001
SUV	-0.2894
VAN	**-0.4207**
FOURWD	-0.0720
CY1989	0.4423
CY1990	0.6609
CY1991	**0.7896**
CY1992	0.4754
CY1993	**0.5622**
CY1994	**0.4935**
CY1995	0.3491
CY1996	0.3362
CY1998	-0.0360
CY1999	0.1595
CY2000	-0.0225
CY2001	-0.1303

Bold entries have chi-square probability less than 0.05

year 1997. First note that the effect of the beam is no longer statistically significant. Because side door beams were not present for a number of years, and were then installed in either 1993 or 1994, calendar year would strongly relate to the presence of beam. Clearly, as years passed, a larger proportion of the light truck fleet would be equipped with the beams, potentially crediting calendar year (or 'time') with some portion of the changes due to the presence of beams.

Note the entry for vehicle age, which is not statistically significant in this model, contrary to the evident trends shown in Exhibits 6 and 8. In addition, several of the individual calendar years are statistically significant, notably the two years in which side door beams were introduced. This differs sharply from the data presented in Exhibit 4, which showed the actual calendar year effects observed in the data. The effects for calendar year in this model are much stronger than those exhibited in the data. For example, the estimate of the effect for calendar year 1993 is a statistically significant 0.5622. This is much larger than the effect of 0.079 shown for 1993 in Exhibit 4. This model gives results contradictory to that which has been seen in earlier analyses. This is clearly not the best fitting model for the data.

An alternative method would be to represent calendar year as a linear variable, rather than as dummy variables. Exhibit 12 presents the results of that regression, using the earliest calendar year (1989) as the year zero.

Exhibit 12: Maximum Likelihood Estimates, Logistic Regression with Linear Calendar Year

Parameter	SV Nearside
Intercept	**-1.3085**
BEAM	-0.0952
AB	**0.3912**
ABS	0.2217
RWAL	-0.2033
RFPASS	**0.6521**
FEMALE	0.1411
AGE	0.0168
AGE_SQD	**-0.0004**
VEHAGE	0.0212
BRANDNEW	-0.0017
BELT	**0.4251**
CURBWT	0.0001
SUV	-0.2827
VAN	**-0.4103**
FOURWD	-0.0761
CALYR	-0.0759

Bold entries have chi-square probability less than 0.05

Note that, although the calendar year variable is not statistically significant, it continues to interact with other variables, resulting in output that is counterintuitive. Again, beam as well as vehicle age are not statistically significant. If the inclusion of calendar year is overpowering the

39

beam variable, additional knowledge could be gained by restricting the data to only those calendar years after the beams were installed. Exhibit 13 presents data for three additional regression models – without calendar year, with calendar year dummy variables, and with calendar year as a linear variable – using only data from calendar year 1994 and later, but still including the full range of model years from 1989 through 1997. Again, only when the calendar year variables are removed is the vehicle age factor statistically significant. As was previously demonstrated, the vehicle age factor is both a stronger factor and more meaningful to this analysis than calendar year.

Exhibit 13: Maximum Likelihood Estimates Using Calendar Year 1994-2001 Data, Single Vehicle Nearside Fatalities

Parameter	No CY Variable	Linear CY Variable	Dummy CY Variable
Intercept	**-1.3076**	**-1.1874**	**-1.7299**
BEAM	**-0.2984**	-0.2024	-0.2066
AB	**0.3582**	**0.3835**	**0.3944**
ABS	0.0640	0.1194	0.1233
RWAL	**-0.3553**	-0.3481	-0.3469
RFPASS	**0.6235**	**0.6280**	**0.6319**
FEMALE	0.1322	0.1332	0.1289
AGE	0.0129	0.0128	0.0130
AGE_SQD	**-0.0004**	**-0.0004**	**-0.0004**
VEHAGE	**-0.0427**	-0.0051	-0.0062
BRANDNEW	0.1758	0.1646	0.1471
BELT	**0.4573**	**0.4624**	**0.4695**
CURBWT	0.0001	0.0001	0.0001
SUV	-0.2820	-0.3133	-0.3182
VAN	**-0.4940**	**-0.5119**	**-0.5202**
FOURWD	-0.0517	-0.0554	-0.0515
CALYR	-	-0.0447	-1.7299
CY1989	-	-	-0.2066
CY1990	-	-	0.3944
CY1991	-	-	0.1233
CY1992	-	-	-0.3469
CY1993	-	-	0.6319
CY1994	-	-	0.1289
CY1995	-	-	0.0130
CY1996	-	-	-0.0004
CY1998	-	-	-0.0062
CY1999	-	-	0.1471
CY2000	-	-	0.4695
CY2001	-	-	0.0001

Bold entries have chi-square probability less than 0.05

When calendar year is included in the model, the obtained estimates for the calendar year variable are stronger than is warranted by the actual data. This inflated importance is due to its colinearity with other, more important, variables in the model, particularly vehicle age and presence of the side door beam. Because of this, the regression model without a calendar year factor provides more meaningful results, and therefore would be considered the best model.

Best Estimates of Effectiveness

Overall, when changes in the fatality rate in side impacts were compared to those in frontal impacts (Exhibits 2a-c), drivers alone or in combination with right front passengers saw the greatest improvement in single vehicle nearside crashes. Improvements in effectiveness for drivers alone ranged from 15 to 21 percent, while in combination with right front passengers the effectiveness ranged from 7 to 14 percent. Effectiveness estimates for drivers in single vehicle far side crashes ranged from 2 to 7 percent. Single vehicle side impacts in general saw improvements, ranging from 9 to 14 percent for drivers alone, and 5 to 13 percent in combination with right front passengers. Right front passengers alone generally experienced the greatest effectiveness in multi-vehicle far side crashes, ranging from 21 to 35 percent. For one year before and after beam installation, the effectiveness was 37 percent, but improvements were not seen with the increased data. Far side crashes in general experienced high effectiveness for right front passengers, ranging from 17 to 27 percent. While several of these effectiveness rates are quite high, none of them were found to be statistically significant.

Several statistical tests on the exposure data (Exhibits 3a-c) did, however, show statistically significant results. For drivers alone, nearside fatalities showed effectiveness rates ranging from 12 to 15 percent, three of the four sets of data (using one to four years of data before and after initial beam installation) being statistically significant. Effectiveness in preventing single vehicle nearside driver fatalities was statistically significant in all sets of data, ranging from 22 to 30 percent. Far side fatalities improved for right front passengers, with effectiveness estimates ranging from 23 to 27 percent and showing significance when three or four years of pre/post beam installation data were used. Multi-vehicle far side fatalities also improved for right front passengers, with effectiveness rates ranging from 25 to 35 percent, statistically significant when more than one year of before and after data was used.

The regression analyses (Exhibits 8 through 10) produce similar but more precise effectiveness estimates. As can be seen by comparing the estimates for "BEAM" in Exhibit 8, as well as Exhibits 9 and 10, the strength of the beam variable is similar to what was seen in the previous analyses of effectiveness and exposure. The addition of side door beams is associated with a statistically significant lowering of side impact fatalities relative to frontal impacts for drivers alone as well as combined with right front passengers in all single vehicle side as well as single vehicle nearside impacts. Right front passengers alone saw no statistically significant effect due to the beam for any type of crash, but the point-estimates for single-vehicle nearside impacts were similar to those for drivers. There were no statistically significant effects for the addition of the beam in any of the multi-vehicle crashes.

The regression analysis allowed several factors to be investigated at the same time, such as seat position, presence of air bag, and vehicle age, which could have affected the previous two analyses. While the first two analyses are more easily understood in terms of changing numbers of fatalities or fatality rates, the regression analysis is the most important, since it combines more information. In addition, the regression procedure allows a larger sample size, since data before and after beam installation do not have to be matched for such situations as presence of air bags or years of production. The regression analysis compares how each factor influences the change in side impacts relative to frontal impacts.

Because the regression model is able to control for important factors influencing crash rates, as well as permitting a larger sample size than the other analyses, the resulting effectiveness estimate would be considered the best, in terms of representing fatalities prevented due to the installation of side door beams. Two "best" effectiveness estimates will be considered. First, an overall reduction of 19 percent for all single-vehicle side impacts, including near- and far side. In addition, a more conservative estimate of 25 percent reduction of nearside single-vehicle fatalities only, with no effect assumed in the far side crashes.

Confidence bounds can be calculated for these estimates using the formula

$$1 - e^{b \pm 1.96 * s}$$

where b is the estimate for beam in the regression equation and s is the standard error of the estimate. For single vehicle nearside impacts, the value is calculated using –0.2901 as the estimate for beam and 0.1076 as the standard error, as shown in Exhibit 8. Thus, the beams are 25 percent effective in preventing front outboard fatalities, with a confidence band of 8 to 39 percent. Little effect was found in multi-vehicle crashes. It is possible that the beam is also effective in single vehicle far side crashes. While data here showed no statistical significance, typically they showed positive results. The best effectiveness estimate of reducing fatalities in these crashes would be 11 percent, with a lower confidence bound of –13 percent and an upper limit of 30 percent. This band contains zero effect, and even includes a negative effect, reflected in the fact that this value was not statistically significant. Overall, fatalities to front outboard occupants were reduced about 19 percent in all single vehicle side impacts, with a confidence band of 4 to 32 percent (using –0.2150 as the beam estimate and a standard error of 0.0877).

Lives Saved by Side Door Beams

Using 19 percent as the estimate of the effectiveness of preventing side impact front outboard occupant fatalities in single vehicle crashes, the annual number of lives saved by side door beams can be determined. During calendar years 1999 through 2002, there were 1,210 front outboard fatalities in single vehicle side impacts in vehicles without a side door beam, and 1,594 fatalities in those with the beams. This is an annual average of 303 fatalities without the beams and 399 with them. Since some fatalities were prevented by vehicles having the beams, this number needs to first be adjusted to determine the potential number of fatalities had none of the vehicles been equipped with beams. Using the formula

$$N_{b_0} + \frac{N_{b_1}}{(1 - e)}$$

where N_{b_0} is the number of fatalities in vehicles without side door beams, N_{b_1} is the number of fatalities in vehicles with side door beams and e is the effectiveness estimate (0.19 in this case), the number of fatalities had no vehicles been equipped with beams is determined to be 796. This was calculated as

$$303 + \frac{399}{1 - 0.19} = 796$$

Given the current mix of vehicles with and without beams, an average of 94 lives were saved annually over the years 1999 through 2002, calculated from the 796 potential fatalities minus the 702 (303 without beams and 399 with them) actual fatalities. This is based on the average effectiveness of beams in both far and nearside impacts, which is then applied on the total of both far and nearside fatalities. Nineteen percent of the 796 potential fatalities could have been prevented if all light trucks on the road had been equipped with side door beams. Thus, 151 lives would be saved annually because of the beams. Using the confidence bands determined in the previous section, a 95 percent confidence interval can be determined for this estimate of lives saved. Using the formula above, but substituting the lower bound of 4 percent (in place of the 19 percent estimate) results in a lower limit of 29 lives saved, calculated as

$$0.04 \times \left(303 + \frac{399}{1 - 0.04} \right) = 29$$

Using the upper bound of 32 percent provides an upper limit of 285 lives saved, calculated as

$$0.32 \times \left(303 + \frac{399}{1 - 0.32} \right) = 285$$

Looking only at single vehicle nearside impacts, the above formula can be used with the effectiveness estimate of 25 percent. Since there was an annual average of 242 front outboard nearside fatalities with beams and 172 without, this is calculated as:

$$172 + \frac{242}{1 - 0.25} = 495$$

Of these 495 potential fatalities, 25 percent could have been saved if all vehicles had side door beams, resulting in 124 lives saved. Again, 95 percent confidence bands can be determined for this estimate also, calculated as above using a lower bound of 8 percent and an upper bound of 39 percent. These are calculated as

$$0.08 \times \left(172 + \frac{242}{1 - 0.08} = 495 \right) = 35$$

and

$$0.39 \times \left(172 + \frac{242}{1 - 0.39} \right) = 222$$

This results in a confidence band around the estimate of 124 ranging from 35 to 222 lives saved in single vehicle nearside impacts.

References

Final Regulatory Impact Analysis: New Requirements for Passenger Cars to Meet a Dynamic Side Impact Test FMVSS 214. U.S. Department of Transportation, National Highway Traffic Safety Administration, Office of Regulatory Analysis, Washington, D.C., August, 1990.

Final Regulatory Impact Analysis Extension of FMVSS 214 Quasi-static Test Requirements to Truck, Buses, and Multi-purpose Passenger Vehicles with a Gross Vehicle Weight Rating of 10,000 Pounds or Less. U.S. Department of Transportation, National Highway Traffic Safety Administration, Office of Regulatory Analysis, Washington, D.C., May, 1991.

Kahane, C.J. *An Evaluation of Side Structure Improvements in Response to Federal Motor Vehicle Safety Standard 214.* NHTSA Technical Report Number HS 806 314. U.S. Department of Transportation, National Highway Traffic Safety Administration, Washington, D.C., November, 1982.

Kahane, C.J. *An Evaluation of Standard 214.* NHTSA Technical Report Number HS 804 858. National Technical Information Service, Springfield, Virginia, 1979.

Kahane, C.J. *Evaluation of FMVSS 214 Side Impact Protection Dynamic Performance Requirement.* NHTSA Technical Report Number HS 809 004. U.S. Department of Transportation, National Highway Traffic Safety Administration, Washington, D.C., October, 1999.